Chickens

by Cari Meister

Bullfrog
Books

Ideas for Parents and Teachers

Bullfrog Books give children practice reading nonfiction at the earliest levels. Repetition, familiar words, and photos support early readers.

Before Reading

• Discuss the cover photo. What does it tell them?

• Look at the picture glossary together. Read and discuss the words.

Read the Book

• "Walk" through the book and look at the photos. Let the child ask questions. Point out the photo labels.

• Read the book to the child, or have him or her read independently.

After Reading

• Prompt the child to think more. Ask: Would you like to have chickens? Why or why not?

Bullfrog Books are published by Jump!
5357 Penn Avenue South
Minneapolis, MN 55419
www.jumplibrary.com

Library of Congress Cataloging-in-Publication Data
Meister, Cari.
Chickens / by Cari Meister.
    p. cm. — (Bullfrog books: animals on the farm)
Includes index.
Includes bibliographical references and index.
Summary: "A hen narrates this photo-illustrated book describing the body parts and behavior of chickens on a farm. Includes photo glossary."
—Provided by publisher.
ISBN 978-1-62031-000-7 (hardcover : alk. paper)
ISBN 978-1-62031-627-6 (paperback)
1. Chickens—Juvenile literature. 2. Chickens—Behavior—Juvenile literature. I. Title.
SF487.5.M45 2013
636.5—dc23
                                2012008221

Series Editor: Rebecca Glaser
Series Designer: Ellen Huber
Photo Researcher: Heather Dreisbach

Photo Credits: All photos by Shutterstock except the following: Dreamstime, 13, 17, 23d; Getty Images, 18, 23e; iStockphoto, 5, 23f; Veer, 9, 10, 14a

Printed in the United States of America at Corporate Graphics, in North Mankato, Minnesota.

Table of Contents

Chickens on the Farm

I am a chicken.
I live on a farm.
Have you ever
seen a chicken?

5

Do you see my flock?

We are all hens.

Do you see
the rooster?

8

comb

He has a big comb.

Do you see the fluffy chicks?

They need a warm place.

Their feathers do not keep them warm yet.

wattle

Do you see my
meaty wattle?

It helps me
keep cool.

It shakes
when I run.

Do you see
the rooster's
sharp claws?

He scratches in the dirt.

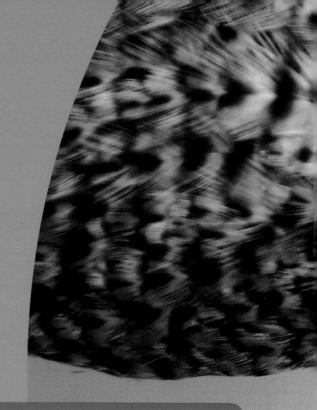

Do you see my beak?
I peck the ground to find food.
I eat worms, seeds, and bugs.

beak

Do you see my
nesting box?
I lay eggs there.

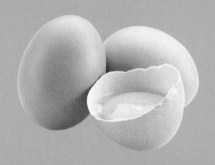

The farmer takes my eggs.
She makes breakfast!

Parts of a Chicken

comb
The brightly colored crest on a chicken's head.

wattle
The fleshy part that hangs down from a chicken's neck; it helps with blood flow, cooling the chicken.

beak
The hard, horny part of a chicken's mouth.

claw
A hard, curved, sharp nail on a bird's foot.

Picture Glossary

chick
A baby chicken.

hen
A female chicken.

feather
One of the light, fluffy parts of a bird's body.

nesting box
A wooden box where farm chickens lay eggs.

flock
A group of chickens that live together.

rooster
A male chicken.

Index

To Learn More

Learning more is as easy as 1, 2, 3.

1) Go to www.factsurfer.com

2) Enter "chicken" into the search box.

3) Click the "Surf" button to see a list of websites.

With factsurfer.com, finding more information is just a click away.